I0221301

James Monteith

First Lessons in Geography

Designed for Beginners

James Monteith

First Lessons in Geography
Designed for Beginners

ISBN/EAN: 9783337166588

Printed in Europe, USA, Canada, Australia, Japan

Cover: Foto ©Paul-Georg Meister /pixelio.de

More available books at **www.hansebooks.com**

MONTEITH'S FIRST LESSONS IN GEOGRAPHY.

VIEWS IN NORTH AMERICA

The Landing of Columbus on the Island of San Salvador (West Indies), October 12th, 1492

2

NATIONAL GEOGRAPHICAL SERIES.

FIRST LESSONS

IN

GEOGRAPHY:

ON THE

PLAN OF OBJECT TEACHING.

DESIGNED FOR BEGINNERS.

BY JAMES MONTEITH,

AUTHOR OF A SERIES OF GEOGRAPHIES, MAPS, ATLASES, AND A POPULAR SCIENCE READER.

NEW YORK ·:· CINCINNATI ·:· CHICAGO

AMERICAN BOOK COMPANY

CONTENTS.

MAPS.

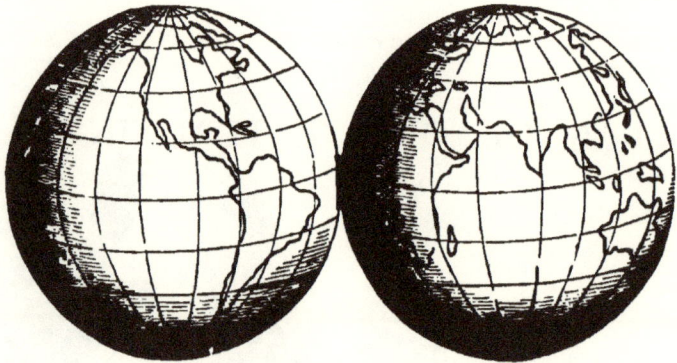

OBJECT LESSONS.

Copyright, 1884, 1889, by

PREFACE.

The plan of this little work is such, that the subject is presented in the most simple form.

It treats of GENERAL FEATURES, such as the locality and description of Continents, Countries, States, Rivers, Mountains, &c., without dwelling prematurely upon details which embarrass the learner in his first effort.

The MAPS are free from all meridians, parallels of latitude, and any superabundance of names; thereby giving a greater prominence to the general divisions of land and water.

The plan of **Object Teaching,** by which the mind receives impressions through the medium of the eye, is here so combined with the Map Exercises, that a child just able to read is at once interested and instructed.

BARNES'S GEOGRAPHIES.

EDITED BY JAMES MONTEITH.

BARNES'S ELEMENTARY GEOGRAPHY...$0.55
BARNES'S COMPLETE GEOGRAPHY.. 1.25
A complete and thoroughly modern two-book series.

OTHER TEXT-BOOKS BY JAMES MONTEITH.

NATIONAL SERIES OF GEOGRAPHIES:
FIRST LESSONS IN GEOGRAPHY.. $0.25
INTRODUCTION TO GEOGRAPHY.. .40
MANUAL OF GEOGRAPHY .... .75

INDEPENDENT SERIES OF GEOGRAPHIES:
ELEMENTARY GEOGRAPHY .... .55
COMPREHENSIVE GEOGRAPHY... 1.10

MONTEITH'S PHYSICAL GEOGRAPHY...................................... 1.00
MONTEITH'S PHYSICAL AND INTERMEDIATE GEOGRAPHY............ 1.20
MONTEITH'S PHYSICAL AND POLITICAL GEOGRAPHY........... ..... 1.15
MONTEITH'S POPULAR SCIENCE READER................................. .75

5

In the Beginning

God Created the Heaven and the Earth

FIRST LESSONS

IN

GEOGRAPHY.

LESSON I.

THIS is a picture of the WORLD or EARTH upon which we live. It is a GREAT BALL. The part you see is the outside or surface, and is either land or water.

The parts of this picture which appear smooth and light represent the WATER; the rough and dark parts, the LAND; and you may know that on the Earth's surface, there is MORE WATER than land.

Look at the picture, and you will observe ships sailing on the water. That great body of water is called an OCEAN, and ships can sail on any part of it, and in any direction. The spots you see in the Ocean are portions of land, called ISLANDS, because there is water all around them.

If you look at the land, on this picture, you will notice black, rough places in it : these are the highest part of the land, and are called MOUNTAINS. The low places between the mountains are called VALLEYS.

On the land, you will see white places; these represent LAKES, which are bodies of water surrounded by the land; and RIVERS, which are streams of water flowing through the land.

The land is higher than the rivers, and the rivers are higher than the ocean; therefore, the rain that falls on the land runs into the rivers, and the rivers flow toward the ocean.

There are FIVE OCEANS on the Earth, named the PACIFIC, ATLANTIC, INDIAN, NORTHERN or ARCTIC, and SOUTHERN or ANTARCTIC. The largest is the Pacific or mild Ocean, which is the one you see before you in the picture.

The water shown at the upper part of the picture is the Northern or Arctic Ocean, and that at the lower part, the Southern or Antarctic.

There are two oceans on the side of the Earth not shown in the picture; they are the Atlantic Ocean, and the Indian Ocean.

Ships sail on all these oceans, as you see them here on the Pacific—just as flies travel around an orange. A fly travels on the outside or SURFACE of an orange; people and ships move on the SURFACE of the Earth. The fields, hills, roads, rivers, and ponds, are parts of the Earth's surface.

Look again at the picture, and you will see land all along the right hand side, stretching from the upper part of the picture almost to the lower, or from the NORTH to the SOUTH. That land is AMERICA. At the middle, or CENTER, the land is very narrow, and because it is a narrow neck of land it is called an ISTHMUS. The land from this isthmus toward the north is NORTH AMERICA, and that from the isthmus toward the south is SOUTH AMERICA. North America and South America are joined to each other by the ISTHMUS OF DARIEN or PANAMA.

Point to North America,—to South America.

On the right hand side of this picture or illustration you may observe a steamship sailing on the Atlantic Ocean. That is but a small part of the Atlantic. From what land does the steamship appear to be sailing? To what land is it sailing? Is it sailing north, or south?

The land which you see on the left hand side of this picture is a part of Asia. The other part of Asia is on the side of the earth opposite that which is here represented; also Europe and Africa. North America, South America, Europe, Asia, and Africa are GRAND DIVISIONS of the land on the Earth.

ABOUT DAY AND NIGHT.

The right hand side of the picture appears light, and the left hand side, dark. That is because the Sun is represented to be shining on the right hand side, where it is DAY; the opposite side being in the shade, has NIGHT.

On the picture, in North and South America, it is day; in Asia, it is night. To cause day in Asia, the Earth must turn around so as to bring Asia toward the Sun. Now, any little boy or girl can tell whether America will have day or night, when Asia has day.

The side of the Earth which is toward the Sun has day, and the opposite side has night; therefore, as the Earth turns around, or whirls like a top, every place will have day, then night, then day again, and so on continually. The Sun shines on one half of the Earth at a time. If the Earth did not whirl about, it would be day continually on the side toward the Sun, and continual night in all countries on the opposite side.

As God caused the Sun to shine upon the Earth to give day, what change would take place with day and night, if He should cause the Sun to cease shining?

Look again at the picture, and you will notice the Sun shining on one side of the Earth, and the Moon shining on the opposite side, where it is night. The world is at that time between the Sun and Moon, which is always the case when you see the Moon full and bright. The STARS you see at night are large shining bodies like the Sun, but appear smaller than the Moon or the Sun, because they are much further from us. The Earth is larger than the Moon, and the Sun is much larger than the Earth. The Moon is nearer to us than the Sun.

On the picture you see CLOUDS. Very few little children know what clouds are, and how rain comes from them. So a few words will be here said about them. You have all seen rising from boiling water, something that appears like smoke. It is not smoke, but VAPOR, to which the water is changed by the heat; and if you would hold a cold basin over that vapor, you would see the vapor turn again to water. In the same way, heat causes vapor to rise from the ocean, lakes, rivers, ponds, etc., and float in the air, until it meets cold air, when it is changed back to water, and returns in the form of drops, and is called RAIN.

So the VAPOR rises from the water; and, while in the form of clouds, the wind blows it over the dry ground, until it is turned into drops, when it comes down to water the grain, the grass, and the flowers; which, by their bright looks and sweet odors, express their joy and thankfulness to God, who alone can do such wonders.

THE EARTH'S SURFACE, SHOWING PARTS OF THE EASTERN AND WESTERN CONTINENTS WITH THE ATLANTIC OCEAN BETWEEN THEM. The Earth is round; therefore the first part of a distant ship which we can see as it approaches us is the top of its mast; next, its sails, and last of all, the ship itself.

What is Geography?
A description of the Earth's surface.

What is the Earth?
The planet or body on which we live.

What is the shape of the Earth?
Round, like a ball.

Of what is the Earth's surface composed?
Land and water.

What is a Continent?
The largest division of the land.

How many Continents are there?
Two; the Eastern and the Western.

On which Continent do we live?
On the Western Continent.

What are the divisions of the Western Continent?

North America and South America.

What are the divisions of the Eastern Continent?

Europe, Asia, and Africa.

What is an Ocean?

The largest division of the water.

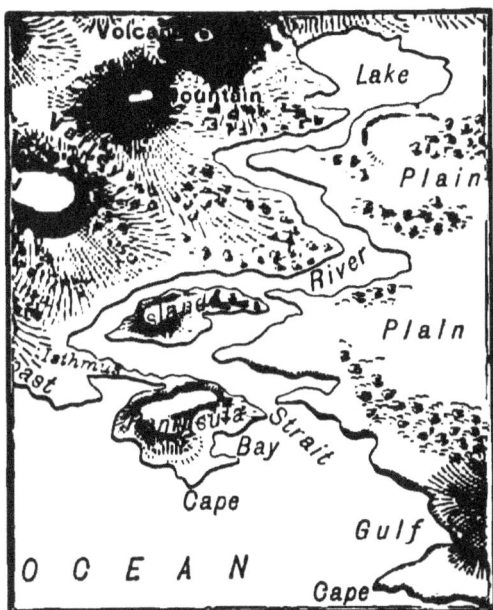

How many Oceans are there?

Five; Pacific, Atlantic, Indian, Southern or Antarctic, Northern or Arctic.

Which is the largest Ocean?

The Pacific Ocean.

What is an Island?

A portion of land *entirely* surrounded by water.

What is a Peninsula?

A portion of land *almost* surrounded by water.

What is a Lake?

A body of water *almost* surrounded by land.

What is an Isthmus?

A neck joining two larger portions of land.

What is a Strait?

A passage connecting two larger bodies of water.

What is a Cape?

A point of land extending into the water.

What is a Mountain?

A vast elevation of land.

What is a Hill?

A small elevation of land.

What is a Volcano?

A mountain which sends out fire, smoke, and lava.

What is a Valley?

The low land between hills or mountains.

What is a Plain?

A level tract of land.

What is a Desert?

A barren region of country.

What is a Sea?

The division of water next in size to an ocean.

What is a Gulf or Bay?

A body of water extending into the land.

What is a River?

A stream of water flowing through the land.

By what are Rivers formed?

By Springs.

What is a Cataract or Waterfall?

Water flowing over a precipice.

The pupils will, upon this illustration, point out the different divisions of land and water.

Point to a MOUNTAIN. Why? *Ans.* Because it is a vast elevation of land. Point to a VOLCANO. Why? *Ans.* Because it is a Mountain which sends out fire, smoke, and lava. Point out the following, and give the reason:—A HILL — VALLEY —ISLAND — PENINSULA—LAKE — ISTHMUS —STRAIT—CAPE—PLAIN— BAY—RIVER. In which of these do you see a church? Show which house stands on a hill—on a mountain. On which division are the cattle grazing?

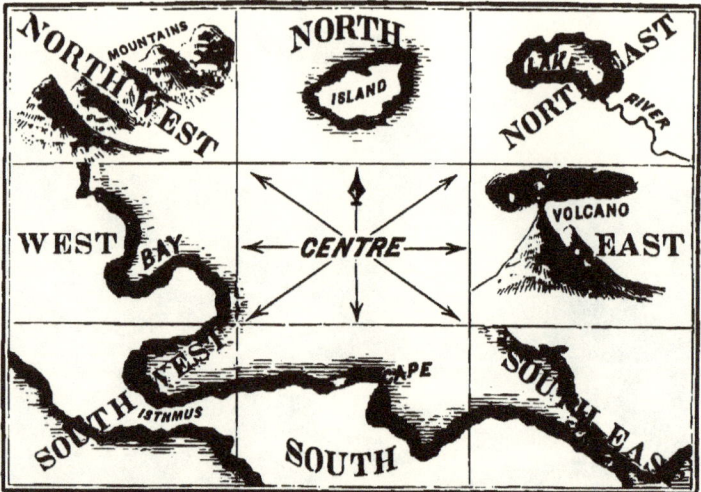

What is a Map?

A picture of the whole, or a part, of the Earth's Surface.

What are the directions on a Map?

Toward the top, North; toward the bottom, South; to the right, East; to the left, West.

In what direction from the center of the picture is the Island?

North.

In what direction is the volcano?   The Cape?

The Bay?   The Lake?   The Strait?   The Mountains?

The Isthmus?

What is in the East?   In the West?   In the South?   In the North?   In the Northwest?   In the Southeast?   In the Northeast?   In the Southwest?

Here is shown a part of a RIVER with a sail-boat
on it. A man is sitting in the stern of the boat, and,
by means of the helm or rudder, he steers it in any
direction. The forward part of a boat is called the
bow. A sloop has one mast; a schooner, two.

On the BANK or edge of this river is a windmill,
with its long arms spread, which, being broad and
light, are blown by the wind round and round, like a
great wheel. In the mill are two large flat stones,
one of which is moved against the other, face to face,
by the arms; so that whatever the miller places between the stones is ground fine like flour.

Corn when ripe and dry is ground into Indian
meal or corn meal; oats when ground we call oat
meal; wheat or rye ground, is flour. Tell how corn
is planted, and when it ripens.

WESTERN HEMISPHERE

ARCTIC OR NORTHERN OCEAN

ASIA

GREENLAND

Baffin B.

Seal

Bering Strait

AMERICA

Hudson Bay

Furs

Lumber

Grain

Gold

New York

Cotton

Gulf of Mexico

Sugar

Oranges & Cigars

WEST INDIES

HAWAIIAN ISLANDS

CAPE VERDE ISLANDS

EQUATOR

ISTHMUS OF PANAMA

Amazon R.

India Rubber

Coffee

SOCIETY ISLANDS

Bananas & Cocoanuts

FRIENDLY ISLANDS

SOUTH AMERICA

Diamonds

Rio Janeiro

Hides

NEW ZEALAND

Strait of Magellan

Cape Horn

PACIFIC OCEAN

ATLANTIC OCEAN

ANTARCTIC OR SOUTHERN OCEAN

III. 96

In what Division of the Earth do we live?

In North America.

What Division south of North America?

South América.

When you look at the rising Sun, what Ocean is before you?

The Atlantic Ocean.

Where does the sun rise?

In the East.

Where, then, is the Atlantic Ocean?

East of America.

When you look at the setting Sun, what Ocean is before you?

The Pacific Ocean.

Where does the Sun set?

In the West.

Where is the Pacific Ocean?

West of America.

What Ocean north of America?

The Northern Ocean.

LESSON XIII.

What Strait connects the Pacific Ocean with the Northern Ocean?

What Ocean south of America?

What Isthmus joins South America to North America?

Which is the most northern Country of North America?

What Bay west of Greenland?

What Mountains in N. America?

What Mountains in S. America?

Do you live in North America or in South America?

What Ocean east of America?

What Ocean west of America?

Where is the Northern Ocean?

Where is the Southern Ocean?

In what Ocean are the Hawaiian Islands? The Cape Verde Islands?

Mention the principal products of North America,—of South America.

EASTERN HEMISPHERE

ARCTIC OR NORTHERN OCEAN

Furs

BRITISH ISLES

Wines & Silk

Figs

EUROPE

Black Sea

Caspian Sea

A S I A

JAPAN ISLANDS

PACIFIC OCEAN

Mediterranean Sea

Fruits

ARABIA

Red Sea

HIMALAYA MTS.

Tea

CHINA

Desert of Sahara

ISTHMUS OF SUEZ

ATLANTIC OCEAN

I N D I A

Rice

Coffee

Gulf of Guinea

A F R I C A

Ivory

EQUATOR

BORNEO

JAVA

Coffee

NEW GUINEA

PACIFIC OCEAN

I N D I A N

O C E A N

Gold

Diamonds

Cape of Good Hope

AUSTRALIA

Gold

Gold Wool

TASMANIA

ANTARCTIC OR SOUTHERN OCEAN

What are the Divisions of the Eastern Continent?

Europe, Asia, and Africa.

Which is the largest?

Asia.

Which is the smallest?

Europe.

Which is furthest to the right, or east?

Asia.

Which is furthest south?

Africa.

What Ocean east of Asia?

Pacific Ocean.

What Ocean south of Asia?

Indian Ocean.

What Ocean west of Africa?

Atlantic Ocean.

What Sea south of Europe?

Mediterranean Sea.

LESSON XV.

What Seas southeast of Europe?

What Sea northeast of Africa?

What Desert in Africa?

What Country in the southeast of Asia?

What Country in the southwest of Asia?

What Islands in the west of Europe?

Which is the largest Island in the World? *Australia.*

What Oceans do you find on the Eastern Hemisphere?

What Division between the Atlantic and Indian Oceans?

What Division west of Asia?

What Division south of Europe?

Between what Divisions is the Mediterranean Sea? Red Sea?

Where is Cape of Good Hope?

What do we get from Asia? Africa? Europe?

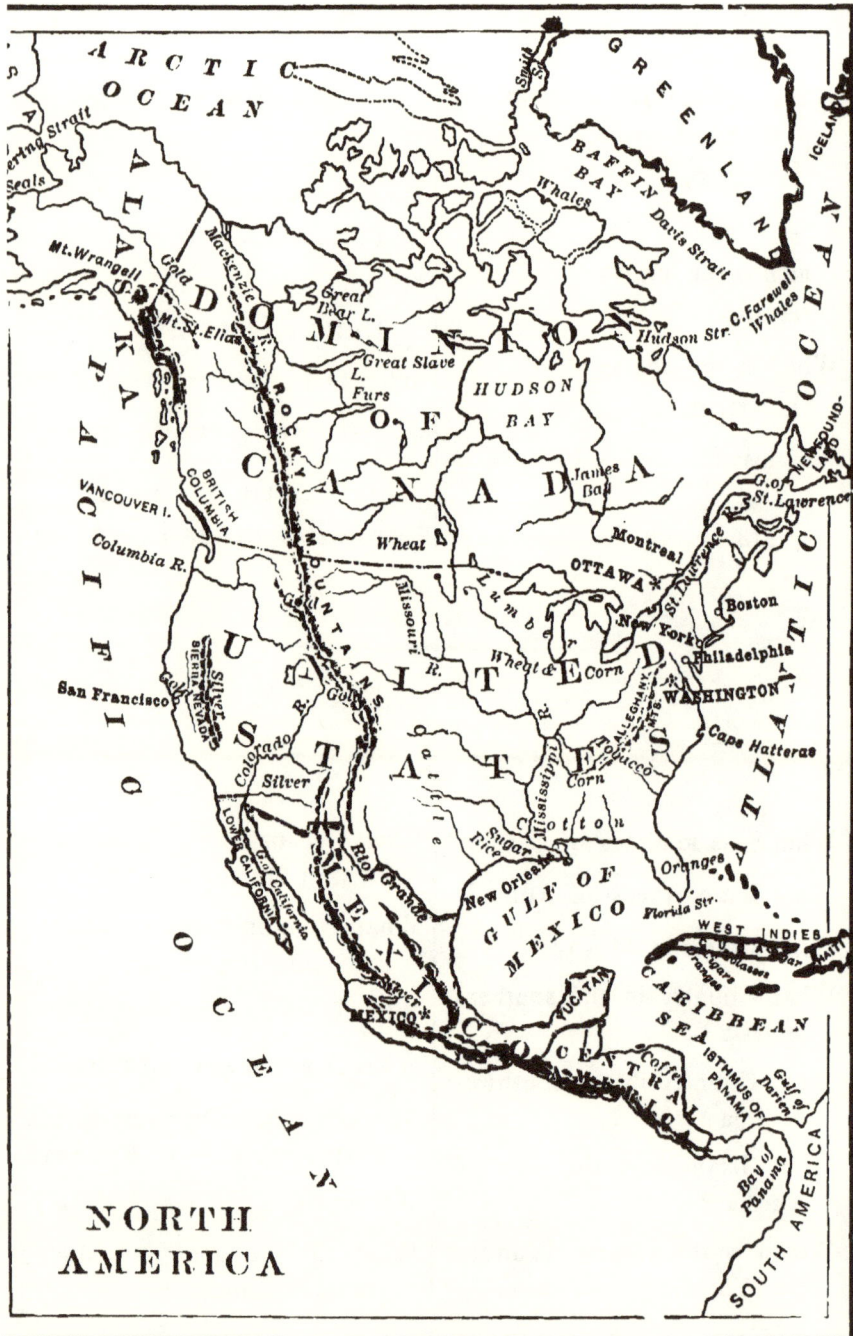

ARCTIC OCEAN

GREENLAND

BAFFIN BAY

Bering Strait

Seals

ALASKA

Mt.Wrangell

Gold

Mackenzie R.

Mt.St.Elias

Great Bear L.

Great Slave L.

Furs

Whales

Davis Strait

ICELAND

Hudson Str.

C.Farewell

Whales

HUDSON BAY

NEWFOUND-LAND

DOMINION OF CANADA

James Bay

G. of St.Lawrence

St.Lawrence R.

VANCOUVER I.

BRITISH COLUMBIA

Columbia R.

Wheat

Lumber

Montreal

OTTAWA

New York

Boston

PACIFIC

ROCKY MOUNTAINS

Missouri R.

Wheat & Corn

UNITED

Philadelphia

WASHINGTON

ATLANTIC

San Francisco

SIERRA NEVADA

Silver

Gold

Colorado R.

Gold

STATES

R.

Cattle

Corn

Tobacco

Cape Hatteras

OCEAN

Silver

Lower California

Rio Grande

Sugar Rice

Cotton

New Orleans

Mississippi R.

Oranges

Gulf of California

MEXICO

MEXICO

GULF OF MEXICO

Florida Str.

WEST INDIES

CUBA

HAITI

Silver

YUCATAN

CARIBBEAN SEA

Molasses

Sugar

Cigars

OCEAN

CENTRAL AMERICA

Coffee

Isthmus of Panama

Gulf of Darien

Bay of Panama

SOUTH AMERICA

NORTH AMERICA

What three Oceans around North America?
Arctic, Atlantic, and Pacific.

What Country furthest north?
Greenland.

What Country furthest south?
Central America.

In what Country do we live?
In the United States.

What Country north of the United States?
The Dominion of Canada

What Country south of the United States?
Mexico.

What Territory in the northwestern part of N. America?
Alaska.

What Peninsula in the south?
Yucatan.

LESSON XVII.

Between what two Oceans is the United States?

What Bay west of Greenland?

What Bay in Canada?

What large Gulf south of the United States?

What large sea southeast of North America?

What Mountains in North America?

What Isthmus south of North America?

Name the Countries of North America, and their products.

Which are the largest two Countries of North America?

Between what two Countries is the United States?

In what Country is Hudson Bay?

Between what two Countries is Baffin Bay?

What large Island southeast of the United States?

Where is Cape Farewell?

MAP-PICTURE OF NORTH AMERICA.

By whom was America discovered?

By Columbus; in the year 1492.

What kind of People did he find here?

Dark-colored Savages.

What did Columbus name them?

Indians.

After whom was America named?

A man named Americus, or Amerigo.

What can you say of the Northern part of N. America?

It is very cold.

What of the Southern part of N. America?

It is very warm.

ICEBERGS NEAR GREENLAND.

This picture represents icebergs, and ships near them. Icebergs are great bodies of solid ice, reaching much higher than the masts of a ship, and extending downward to a great distance below the surface of the water. They will float in the water, and every year some ships are wrecked by coming in contact with them.

Icebergs are most numerous in the Arctic Ocean.

Men have sailed through Baffin Bay, and as far north as the ship you see in the map.

The names of those men who have become famous for their adventures in the frozen regions north of North America, are Sir John Franklin, Dr. Kane, Dr. Hayes, Hall, De Long, Greely, and Schwatka.

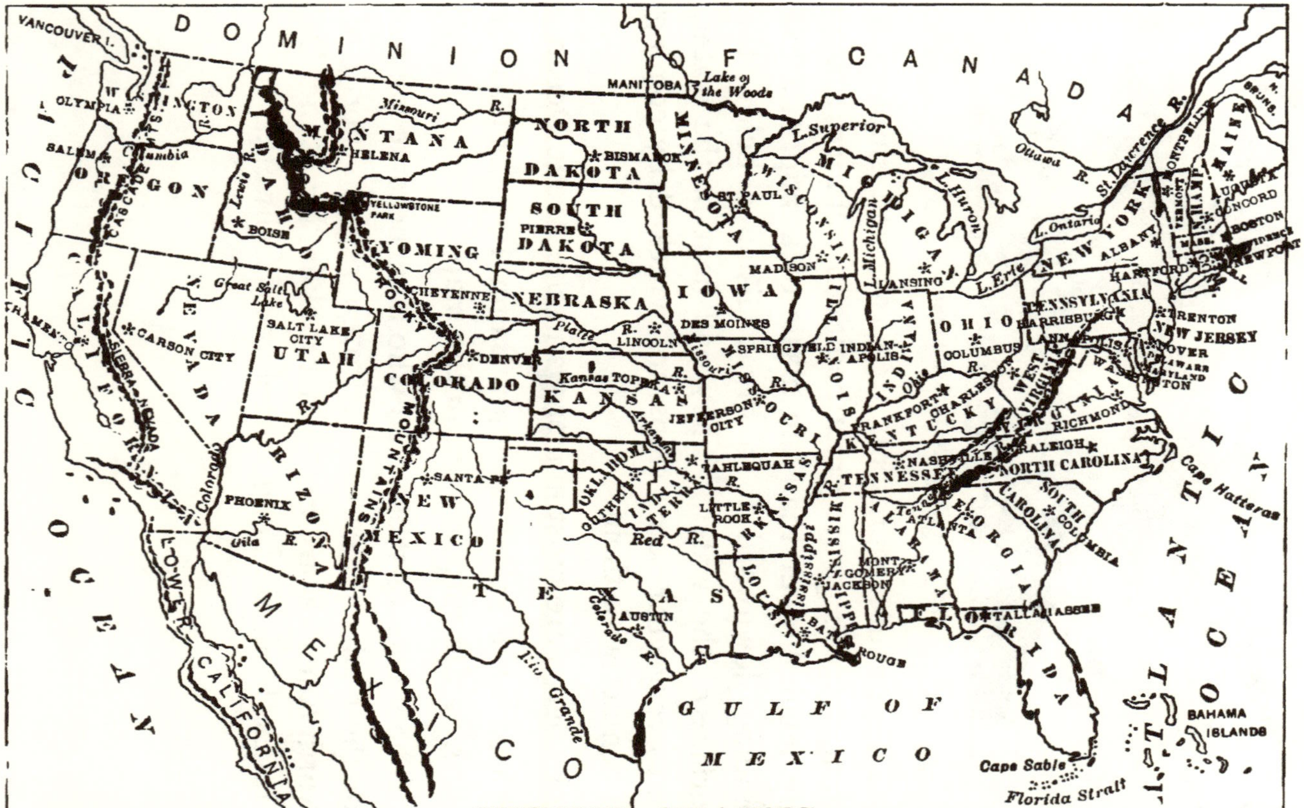

UNITED STATES

LESSON XX.

What is this Country called?

The United States, or the Union.

How many States are there? 45.

How are they divided?

Into Eastern, Middle, Western and Southern States.

Which is the largest State?

Texas.

Which is the smallest State?

Rhode Island.

Which is furthest south?

Florida.

Which is furthest northeast? Maine.

What States border on the Pacific Ocean?

California, Oregon, and Washington.

What State in the north is almost surrounded by Lakes?

Michigan.

Which is the largest of those Lakes?

Lake Superior.

LESSON XXI.

What large River flows south into the Gulf of Mexico?

What large Rivers flow into the Mississippi?

Which is the largest of those Rivers?

What Rivers flow into the Missouri River?

Into what does the Ohio River flow?

What River between Texas and Mexico?

What Mountains extend through the Western part of the United States?

What Mountains nearer the Pacific Coast?

What Ocean east of the United States? West?

What Country north?

What Country and Gulf south?

In what State do you live?

What States surround your State?

Mention the great Lakes.

Where is Lake Superior?

What Lake in Utah? North of Minnesota?

27

LESSON XXII.

What State is furthest northeast? South? West?

Between what Ocean and Gulf is Florida?

What three States touch Lake Superior?

What four States touch Lake Michigan?

What four States touch Lake Erie?

What States border on the Pacific Ocean?

What States border on the Gulf of Mexico?

What States border on the Atlantic Ocean?

What States on the east side of Mississippi River?

What States on the west side?

What three States on the north side of the Ohio River?

What two on the south side?

Through what States and what Territory do the Rocky Mountains extend?

LESSON XXIII.

What Country east of Maine?

What Dominion north of New York? *Canada.*

What three States east of New York?

What two south?

What State and lake north of Ohio?

What River south? State east? West?

What State north of I'owa? South?

What State north of Virginia?

What State south? West? Northwest?

What State north of Louisiana?

What State east? West?

What States north of Kentucky?

What State west? East? Northeast? South?

What Cape east of North Carolina?

What Cape south of Florida?

What large Island south of Florida?

What Strait between Florida and Cuba?

THE UNITED STATES.

GEORGE WASHINGTON.

Who governed this country about 120 years ago?

The king of England.

How did the Americans obtain their freedom?

By a war which lasted nearly eight years.

What great man led the American army?

George Washington, who became the first President of the United States.

The UNITED STATES comprises forty-five States, five Terri-
tories, and one District, and also owns Puerto Rico in the West
Indies, and the Hawaiian Islands and other possessions in the
Pacific Ocean.

CAPITAL OF THE UNITED STATES.

| WASHINGTON, | on the | Potomac River. |

EASTERN OR NEW ENGLAND STATES.

States.	Capitals.		Situation.
MAINE,	Augusta,	on the	Kennebec River.
NEW HAMPSHIRE,	Concord,	on the	Merrimac.
VERMONT,	Montpelier,	on the	Winooski or Onion.
MASSACHUSETTS,	Boston,	on	Boston Harbor.
RHODE ISLAND,	{ Providence,	on	Providence Bay.
	and Newport,	on	Narragansett Bay.
CONNECTICUT,*	Hartford,	on the	Connecticut.

LESSON XXVI.

MIDDLE STATES.

States.	Capitals.		Situation.
NEW YORK,	Albany,	on the	Hudson River.
NEW JERSEY,	Trenton,	on the	Delaware.
PENNSYLVANIA,	Harrisburg,	on the	Susquehanna.
DELAWARE,	Dover,	on	Jones Creek.

* kon-net' e-kut.

NOTE.—The capital of a state or a country is the city in which its laws are made.

LESSON XXVII.

SOUTHERN STATES.

States.	Capitals.		Situation.
MARYLAND,	Annapolis,	on the	Severn River.
VIRGINIA,	Richmond,	on the	James.
NORTH CAROLINA,	Raleigh,	near the	Neuse.
SOUTH CAROLINA,	Columbia,	on the	Congaree.
GEORGIA,	Atlanta,	near the	Chattahoochee.
FLORIDA,	Tallahassee,		Inland.
ALABAMA,	Montgomery,	on the	Alabama.
MISSISSIPPI,	Jackson,	on the	Pearl.
LOUISIANA,	Baton Rouge,	on the	Mississippi.
TEXAS,	Austin,	on the	Colorado.
WEST VIRGINIA,	Charleston,	on the	Kanawha.

LESSON XXVIII.

WESTERN STATES.

ARKANSAS,	Little Rock,	on the	Arkansas River.
TENNESSEE,	Nashville,	on the	Cumberland.
KENTUCKY,	Frankfort,	on the	Kentucky.
OHIO,	Columbus,	on the	Scioto.
MICHIGAN,	Lansing,	on the	Grand.
INDIANA,	Indianapolis,	on the	W. Fork of White R.
ILLINOIS (oy),	Springfield,	near the	Sangamon.
WISCONSIN,	Madison,	on	Fourth Lake.
IOWA,	Des Moines,	on the	Des Moines.
MISSOURI,	Jefferson City,	on the	Missouri.
CALIFORNIA,	Sacramento,	on the	Sacramento.
MINNESOTA,	St. Paul,	on the	Mississippi.
OREGON,	Salem,	on the	Willamette.
KANSAS,	Topeka,	on the	Kansas River.
NEVADA,	Carson City,	on the	Carson River.
NEBRASKA,	Lincoln,	on	Salt Creek.
COLORADO,	Denver,	on	Cherry Creek.
NORTH DAKOTA,	Bismarck,	on	Missouri River.
SOUTH DAKOTA,	Pierre,	on	Missouri River.
MONTANA (mon tä' nä),	Hel' e na,	near	Missouri River.
WASHINGTON,	Olympia,	on	Puget Sound.
IDAHO,	Boise,	on	Boise River.
WYOMING,	Cheyenne,	on	Crow Creek.
UTAH,	Salt Lake City,	on	Jordan River.

More than 200 years ago, this country, now called the United States, was a wilderness, inhabited by Indians, who subsisted upon fish, and the flesh of wild animals which they killed in hunting, and who lived in huts made of bark and the skins of animals.

No cities were built until the country was settled by white men, who came from Europe; and, probably, where your house now stands, Indians once chased the buffalo, bear, or some other wild animal.

Many of the white settlers of this country suffered great cruelties from the Indians, who burned their houses and murdered men, women, and children, as you see in the picture. The Indians now live mostly in the Indian Territory.

The first inhabitants of a place are called settlers, or colonists.

The people of the United States are famous for perseverance and inventive genius. Some years ago, people rode in stage-coaches over rough and hilly roads ; but now they travel by steamboat or railroad.

A STEAMBOAT is moved along by two large paddle-wheels revolving in the water. The wheels are moved by STEAM, which rises from boiling water. Traveling by steamboat began about eighty years ago ; and by railroad, more than sixty years ago.

A STEAMSHIP differs from a Steamboat in having sails besides the steam power. A SAILING VESSEL is moved by the wind blowing against the sails.

The TELEGRAPH you see in the picture is a long iron wire supported by tall poles. At each end of the wire there is an instrument, by which men send messages with lightning velocity. The telegraph was invented by Morse ; the steamboat, by Fulton.

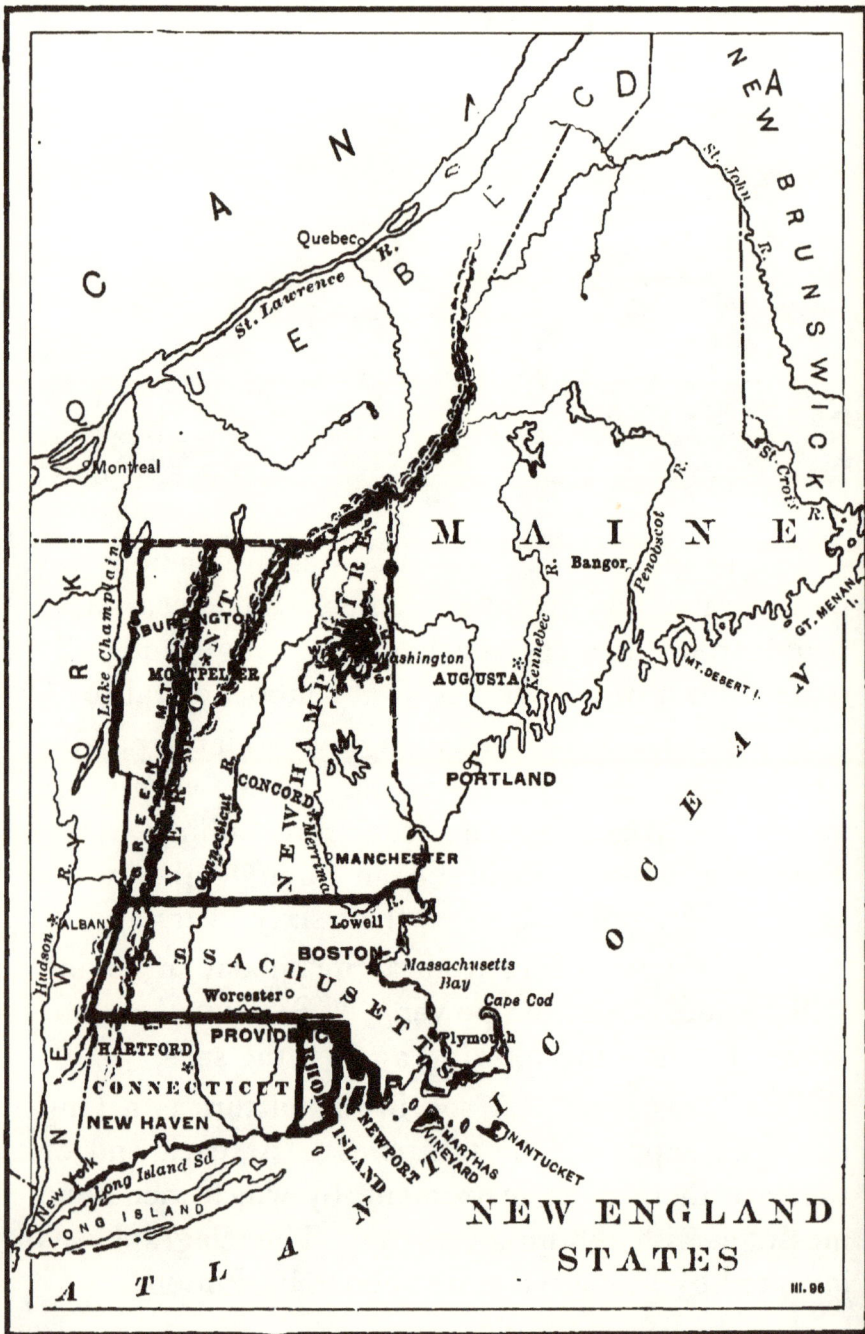

NEW ENGLAND
STATES

Ill. 96

How many Eastern or New England States are there? *Six.*

What two States north of Massachusetts?
New Hampshire and Vermont.

What two States south of Massachusetts?
Connecticut and Rhode Island.

What State west?
New York.

What large River between Vermont and New Hampshire?
Connecticut River.

Through what States does it flow?
Massachusetts and Connecticut.

What Rivers in Maine?
Kennebec and Penobscot.

What River in New Hampshire?
Merrimac River.

LESSON XXXII

Which of the Eastern States touch the Atlantic Ocean?

Which touch New York?

What Lake between Vermont and New York?

What Country north of the Eastern States?

What Country east of Maine?

What large Island south of Connecticut?

What water between Connecticut and Long Island?

Name the Eastern States.

Which is the largest?

Which is the smallest?

Where are the Green Mountains?

Where are the White Mountains?

What large River flows into Long Island Sound?

What Cape in the eastern part of Massachusetts?

What three Rivers flow into the Atlantic Ocean?

· J

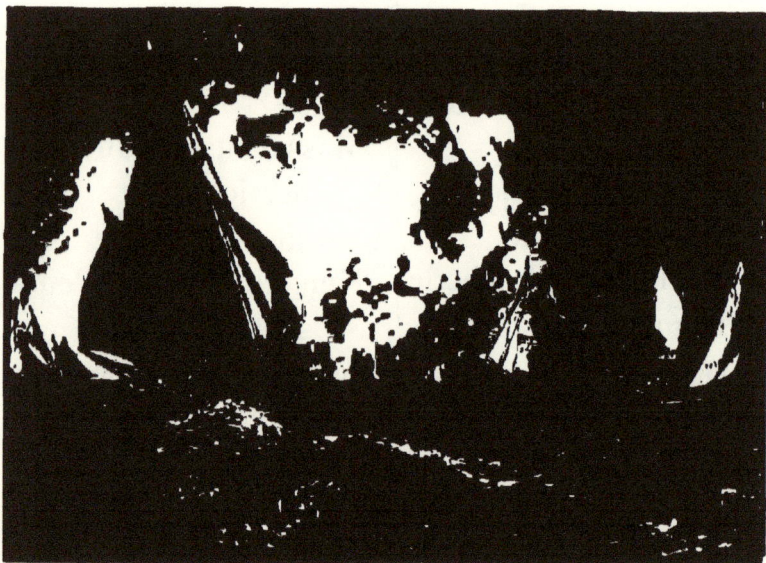

MANY VESSELS ARE ENGAGED IN FISHING OFF THE COASTS OF NEW ENGLAND AND NEWFOUNDLAND.

What are these six States called?

New England.

Who first came to New England?

People from England, called Puritans, also Pilgrims.

In what does Massachusetts excel every other State?

In the manufacture of cotton and woolen goods, and of boots and shoes.

Where was the first cotton-mill in the United States built?

In Rhode Island.

For what is Connecticut noted?

For the manufacture of woolen and cotton goods, iron and wooden wares, clocks and buttons.

This is a view of a canal and a manufactory. In the foreground are sheep and cattle, which are raised in Vermont in large numbers.

COTTON is a soft, white substance obtained from the cotton-plant, which grows in the Southern States. It is brought in bales to the manufactories of the Eastern States; where, by means of machinery, it is drawn out and twisted into threads, and then it is woven into cloth. It can be dyed or printed in colors. Muslins and calicoes for ladies' dresses are made of cotton.

WOOL grows upon sheep, and is cut in warm weather. It is made into threads by spinning, then woven. Blankets and winter clothing are made of wool. So, cotton is obtained from a plant; wool, from an animal.

A CANAL is like a great ditch filled with water, so that boats may be drawn along by horses or mules which walk on a TOW-PATH at the side of the canal. . ∕

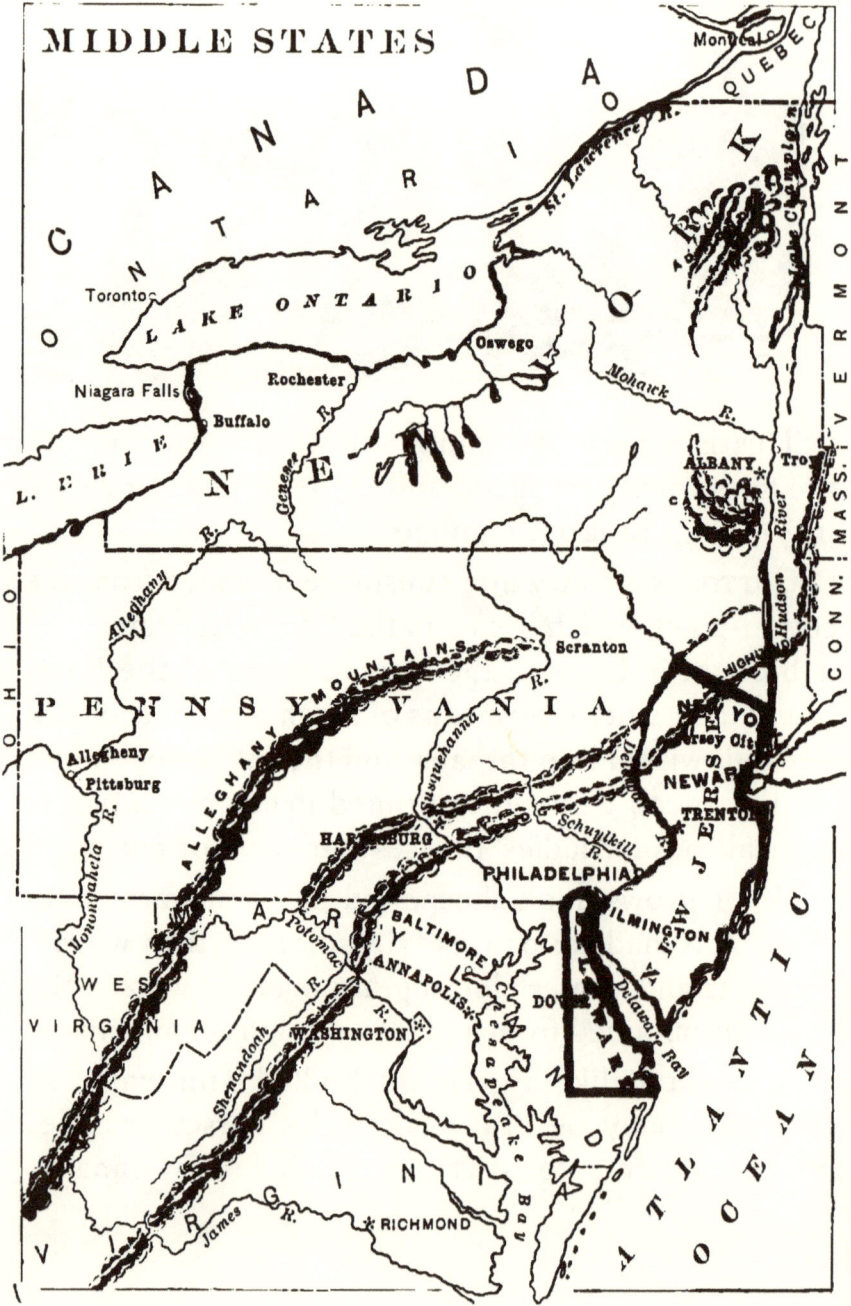

MIDDLE STATES

How many Middle States are there ? *Four.*

Which is the largest ?

New York.

Which is next in size ?

Pennsylvania.

Which is next ?

New Jersey.

Which is the smallest ?

Delaware.

What Country north of New York ?

Canada.

What two Lakes on the West ?

Ontario and Erie.

What large River in the eastern part of New York ?

Hudson River.

What large River in the western part of New York ?

Genesee River.

LESSON XXXVI.

What River between Pennsylvania and New Jersey ?

What large River flows through the eastern part of Pennsylvania ?

What two Rivers meet in the western part ?

What large River flows northeast from Lake Ontario ?

What Mountains in Pennsylvania ?

What Mountains in New York ?

What Bay between New Jersey and Delaware ?

Name the Middle States.

What States south of New York ?

What States east of New York ?

Where is Lake Erie ?

Where is Lake Ontario ?

Where is the Hudson River ?

Where is the St. Lawrence ?

Into what Lake does the Genesee River flow ?

What Lake northeast of New York ?

A TRAIN OF CARS.—A CANAL.

For what are the Middle States noted ?
  For Canals and Railroads.

What can you say of New York ?
  It has more inhabitants than any other State

For what is Pennsylvania celebrated ?
  For Coal and Iron.

What does New Jersey produce ?
  Fine fruits and vegetables.

What does Delaware produce ?
  Excellent wheat, Indian corn and peaches.

What does the word Pennsylvania mean ?
  Penn's Woods.

WILLIAM PENN, a Quaker, came from England to this country, more than two hundred years ago, with many other Quakers, and formed a colony or settlement in Pennsylvania.

Penn was very wise and kind in his dealings with the Indians, and paid them for all the land which his people occupied; consequently, the Indians respected and loved the Quakers very much; and Pennsylvania was the only American colony formed without bloodshed. It has now more inhabitants than any other State in the Union, except New York.

On the left of the picture are Indian women, called SQUAWS, carrying their PAPPOOSES, or babies, which hang on their backs like soldiers' knapsacks.

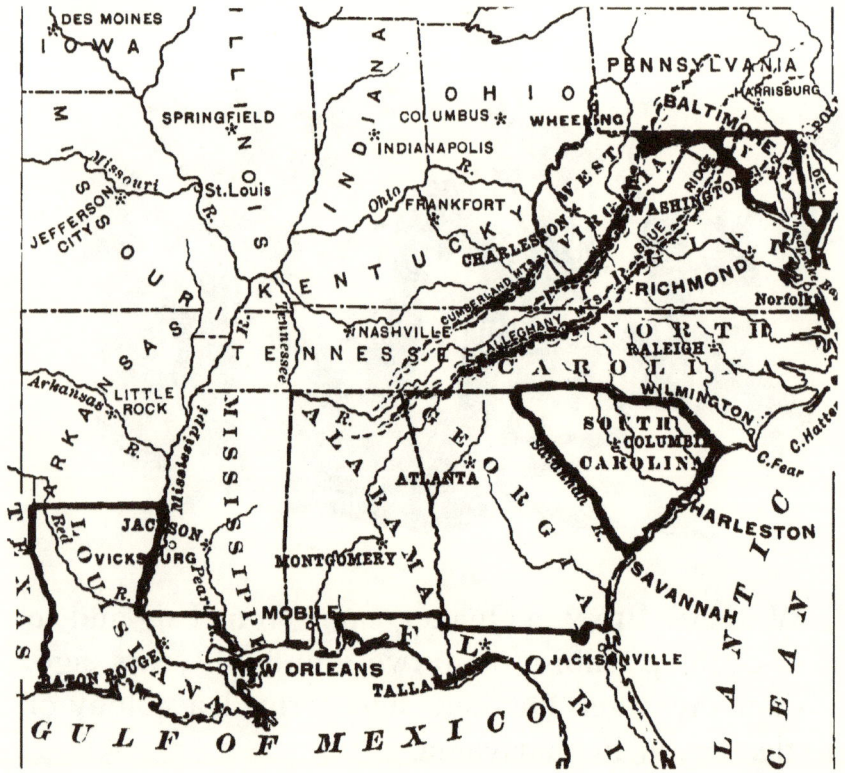

How many Southern States are there? *Eleven.*

Which is the largest?
Texas.

Which is furthest south?
Florida.

What division of land is Florida?
A Peninsula.

What Island south of Florida?
Cuba.

What States north of Florida?
Georgia and Alabama.

What Bay east of Virginia?   .
Chesapeake Bay.

What River flows through the northern part of Alabama?
Tennessee River.

Into what River does the Tennessee flow?
Into the Ohio River.

LESSON XL.

What River between South Carolina and Georgia?

Into what Ocean and Gulf do the Rivers of the Southern States flow?

In what Mountains do most of them rise?

What three ranges of Mountains on this map?

Which are the most mountainous of the Southern States?

What two Western States between the Southern States and the Ohio River?

Name the Southern States.

Which of them touch the Atlantic Ocean?

Name the States that touch the Gulf of Mexico.

What two Southern States touch the Mississippi River?

Between what Ocean and Gulf is Florida?

Where is Cape Sable?

THE SEA SHORE. The waves of the ocean constantly roll against the soft, sandy beach, and form low sand hills which line the coast for many miles.

What is the climate of the Southern States?
Very warm.

What are raised on the plantations of the Southern States?
Cotton, Corn, Sugar-cane, Rice, and Tobacco.

What State excels in the production of Sugar?
Louisiana.

What Presidents were born in Virginia?
Washington, Jefferson, Madison, Monroe, William Henry Harrison, Tyler, and Taylor.

On the left of this picture you see the tall SUGAR-CANE growing ; in the front, COTTON; and on the right, TOBACCO. The sugar-cane is cut and taken to the crushing-mill, where the juice is pressed out, and afterwards boiled,—the sugar settling to the bottom of the kettles, and the MOLASSES remaining at the top.

The leaves of the tobacco-plant are dried before they are ready for use. (See page 67.)

Cotton is a plant which is extensively cultivated in the Southern States. It is formed in a kind of nut-shell, which bursts, and the cotton appears. It is then picked from the covering, and taken to the mill, to be separated from the seeds contained inside. It is afterwards spun into threads, then woven into cloth. The soft, white substance which you have seen growing on the top of a thistle, resembles cotton on the plant.

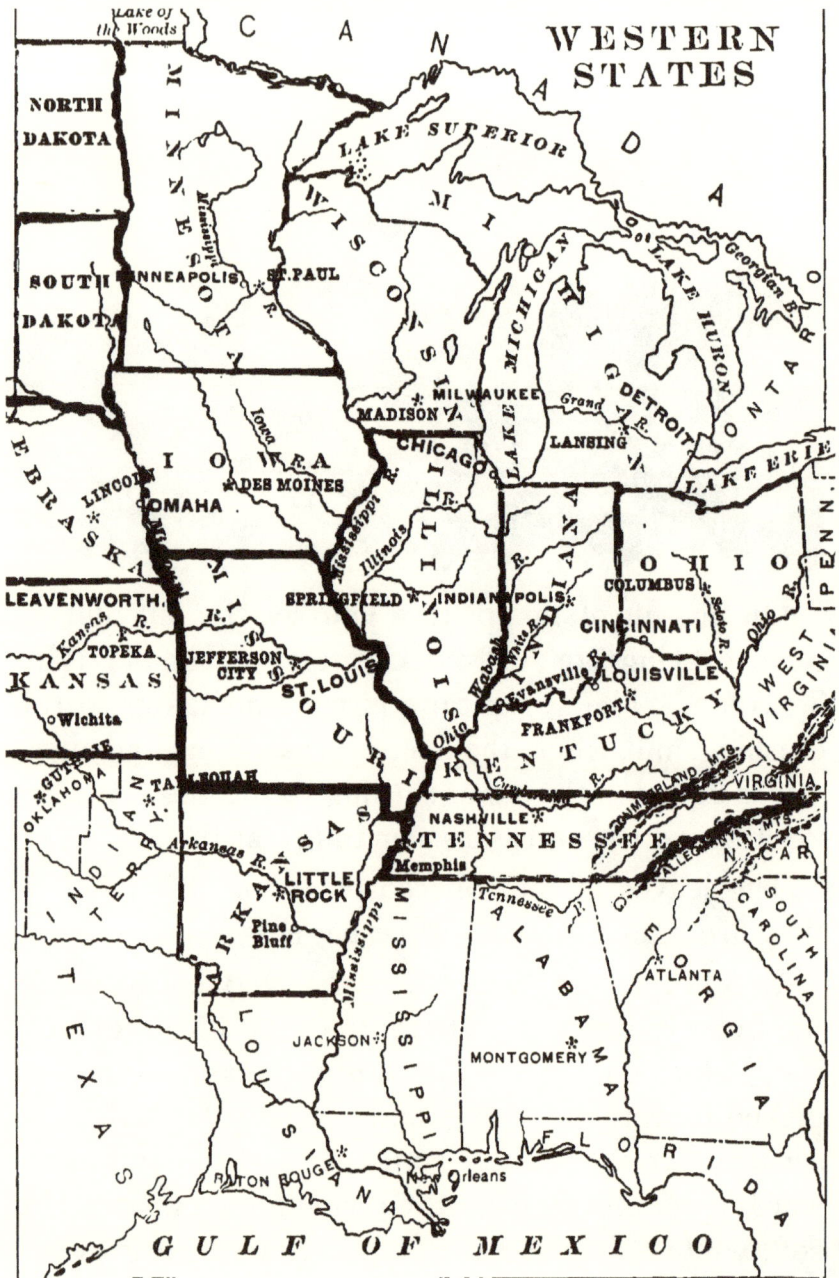

WESTERN
STATES

What States are furthest north and northwest?

Minnesota, North Dakota, Montana, Idaho, Washington (see p. 26).

What four are on the west side of the Mississippi River?

Minnesota, Iowa, Missouri, and Arkansas.

What three lie on the north side of the Ohio River?

Ohio, Indiana, Illinois.

What two south of the Ohio?

Kentucky and Tennessee.

What Lake north of Michigan?

Lake Superior.

What Lake east of Michigan?

Lake Huron.

What Lake west of Michigan?

Lake Michigan.

What Lake north of Ohio?

Lake Erie.

LESSON XLIV.

Into what River do nearly all the other Rivers of the Western States flow?

Which flow into the western side of the Mississippi?

Which into the eastern side?

In what State does the Mississippi rise?

What rivers flow into the Ohio River?

What Western States do not appear on this map? (See p. 31.)

What States border on Lake Superior? On Lake Michigan? On Lake Erie?

What River in Ohio? In Illinois?

What River between Indiana and Illinois?

What States south of Tennessee?

What States east of Kentucky?

Which is the coldest, or most northern, of the Western States?

Which is the warmest?

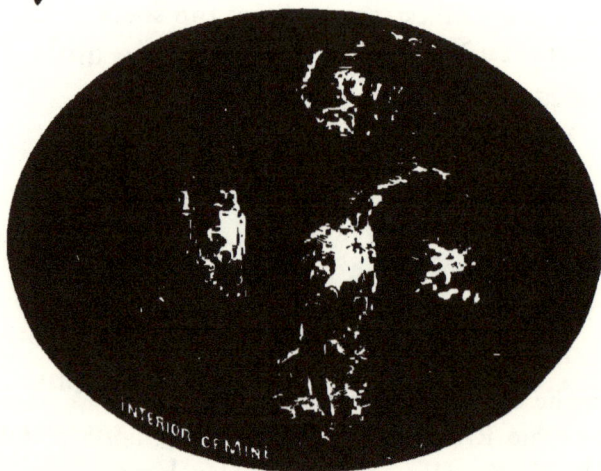

INTERIOR OF MINE

Which is the largest section of the Union?
The Western States.

In what are the people chiefly engaged?
Farming in the East and mining in the West.

What are raised on their extensive farms and fields?
Corn, wheat, oats, rye, and fruit; besides millions of horses, cattle, sheep, and hogs.

What States are noted for gold and silver?
California, Montana, Colorado, Utah, and Nevada.

Which are the most populous of the Western States?
Ohio and Illinois.

What State is noted for corn, wheat, and oats?
Illinois.

A FARMER CUTTING WHEAT.

This is a picture of an overflow of the Mississippi River. Look at your map of the United States, and observe that the greater part of our country is drained by that river; that is, the rain that falls in the States and Territories between the Alleghany and Rocky Mountains, runs into rivers which flow directly or indirectly into the Mississippi. This is caused by the land sloping downward from these two chains of mountains where the land is highest, to the Mississippi River where it is lowest.

To prevent inundations or freshets, which do great damage to houses, farms, and cattle, the people have raised banks, called LEVEES, along the river.

SOUTH AMERICA

What natural division of land is South America?

A Peninsula.

What Division of the Earth is north of South America?

North America.

What Ocean east?

Atlantic Ocean.

What Ocean west?

Pacific Ocean.

What Sea north?

Caribbe'an Sea.

What Bay northwest?

Bay of Panama.

Which is the largest River in South America?

Amazon River.

How long is the Amazon?

Four thousand miles.

What River further north than the Amazon?

Orinoco River.

What Rivers in the southeast?

Parana and La Plata.

LESSON XLVIII.

Into what Ocean do nearly all the Rivers of South America flow?

Which is the largest Country in South America?

Which are further north?

Which further south?

What great chain of Mountains in South America?

Near what Coast do they extend?

What Mountains in Brazil?

What precious stones are found there?

Between what two Oceans is South America?

Where is the Caribbe'an Sea?

What Countries border on that sea?

What Countries border on the Atlantic?

Which on the Pacific?

Which are inland?

What Cape on the North? East? South?

What Strait north of Tierra-del-Fuego?

SOUTH AMERICA.

For what is South America noted?

For the largest rivers and longest mountain-chain in the world.

What precious stones are found in Brazil?

Diamonds.

What has been obtained in Bolivia?

Silver in large quantities.

What animals roam over the vast grassy plains of South America?

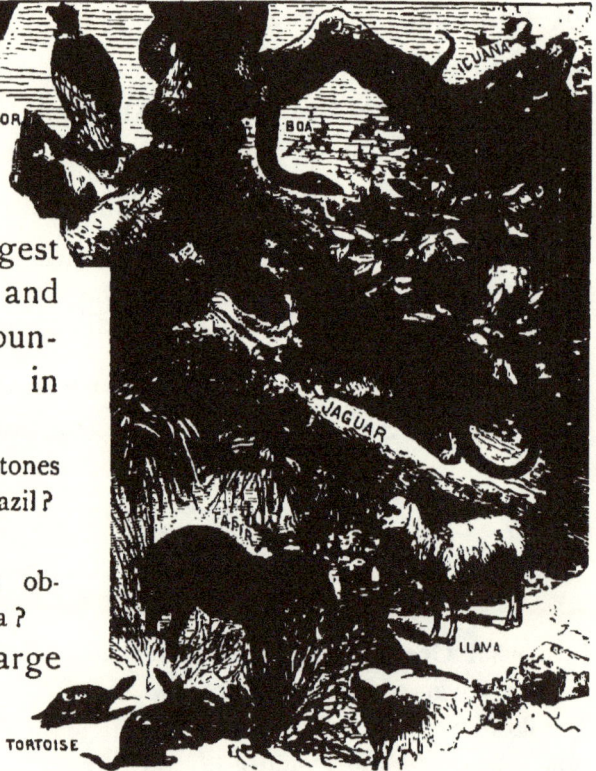

The Condor is the largest bird of flight in the world. The Jaguar is like a leopard; the Tapir, like a large hog; and the Iguana is a kind of lizard. The Llama is a beast of burden.

Horses and cattle in immense numbers.

What remarkable species of serpent in South America?

The Boa Constrictor, which is able to destroy animals as large as deer.

Here is a view of the Andes Mountains, which rise sixty times as high as the highest church steeple you ever saw ; they are so high that their tops are continually covered with snow.   Some are volcanoes.

EARTHQUAKES sometimes occur, and destroy whole cities, killing many of the inhabitants.   An earthquake is a violent shaking of a part of the Earth.

Travelers cross the Andes on the backs of mules, which are much safer than horses in such dangerous places.   Indians, with chairs fastened on their backs, sometimes carry travelers over the mountains with safety.

Immense birds, called CONDORS, are found here, which often destroy sheep and cattle, tearing them with beak and claws.

EUROPE

III. 98

ARCTIC OCEAN

NOVA ZEMBLA

URAL MTS.

North Cape

ICELAND

LAPLAND

Dvina R.

White Sea

SCANDINAVIAN MTS.

SWEDEN

STOCKHOLM

ST.PETERSBURG

Gulf of Bothnia

Volga

R U S S I A

ATLANTIC OCEAN

NORTH SEA

IRELAND

DUBLIN

GREAT BRITAIN

C. Clear

LONDON

BALTIC SEA

DENMARK

Don R.

GERMAN EMPIRE

BERLIN

POLAND

Caspian Sea

FRANCE

PARIS

Danube R.

VIENNA

AUSTRIA

Bay of Biscay

SWITZ.

ALPS

ROUM.

HUNGARY

BLACK SEA

PYRENEES

ITALY

Adriatic Sea

SERVIA

TURKEY

CONSTANTINOPLE

SPAIN

PORTUGAL

LISBON

MADRID

ROME

Naples

GREECE

ARCHIPELAGO

ASIA MINOR

Str. of Gibraltar

MEDITERRANEAN SEA

SICILY

CANDIA

A F R I C A

A S I A

What Ocean north of Europe?
The Arctic Ocean.
What Ocean west of Europe?
The Atlantic Ocean.
What Sea south of Europe?
The Mediterranean Sea.
What is the Mediterranean Sea?
The largest Sea in the world.
What Bay west of France?
The Bay of Biscay.

Which is the largest Country in Europe?
Russia.
Which is the smallest?
Switzerland.
What two countries touch Russia on the southwest?
Austria and Roumania.
What Country south of Turkey?
Greece.

LESSON LII.

What two Countries west of the North Sea?
What Country west of England?
What Country south of England?
What Country south of France?
What Country west of Spain?
Where are the Alps Mountains?
Where are the Pyrenees Mountains?
Where is the coldest part of Europe.
Where is the warmest part?

What two Oceans touch Europe?
What five large Seas do you find on the map of Europe?
What four Rivers?
What Strait connects the Mediterranean Sea with the Atlantic?
What Cape in the northern part of Europe?
What Sea east of Italy?
Where is the White Sea?
Mention all the Countries of Europe.

WILLIAM TELL, a heroic Swiss, in his efforts to obtain liberty for his country, was captured; and, for punishment, was cruelly ordered to shoot an apple placed on the head of his own little son. The arrow cut the apple in two, without injuring the child. This occurred more than 500 years ago.

What can you say of Europe?

It is the smallest Grand Division of the Earth.

Which are the most important divisions of Europe?

England, Germany, Russia, and France.

What are the inhabitants of Europe called?

Europeans.

What does the southern part of Europe produce extensively?

Grapes, Oranges, Lemons, Figs, and Olives.

What important School-law in Prussia and some other countries?

All the boys and girls are compelled to attend school regularly.

Here is a view in the northern part of Europe, which is noted for the great length of its winters, and of its winter nights and summer days.

For several weeks in winter, the people there do not see the sun; but for the same length of time in summer, the sun does not set.

The man you see in the picture is called a Laplander, because he lives in Lapland. Laplanders travel from place to place in sleds drawn by reindeer. The milk and flesh of these animals are used for food, and their skins for clothing. A Laplander's wealth is known by the number of reindeer he owns.

In the southern part of Europe the climate is mild and pleasant; oranges, lemons, figs, olives, grapes, and other fruits being raised in abundance.

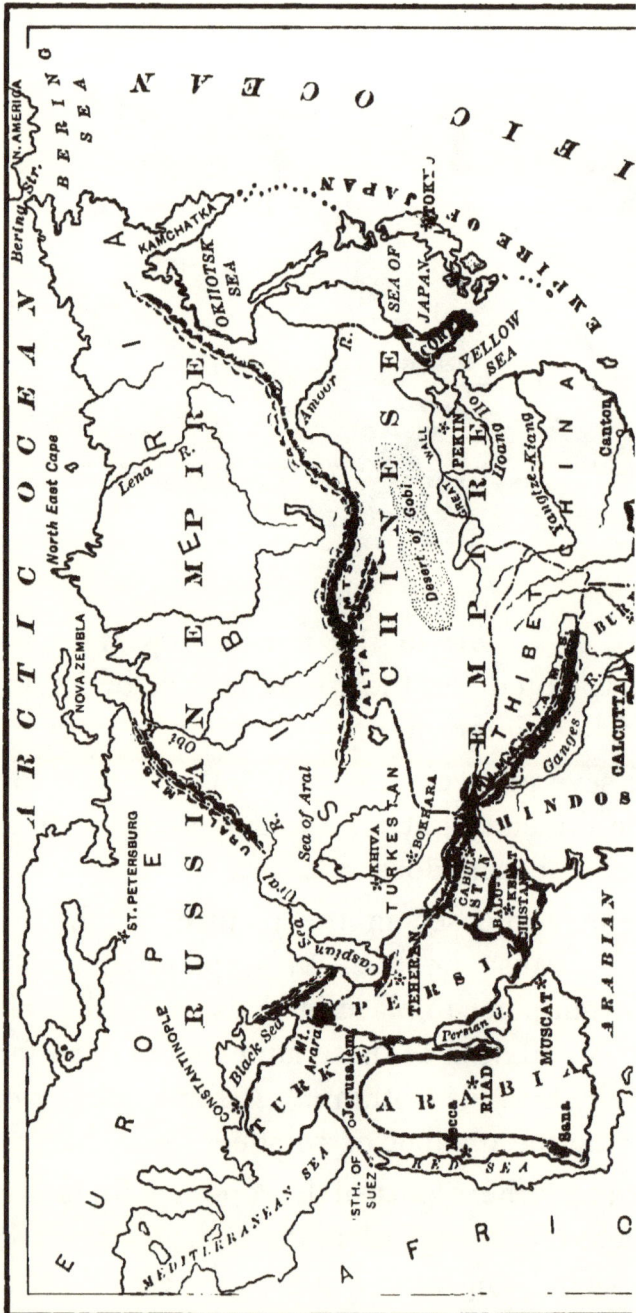

LESSON LV.

What Ocean north of Asia?
Arctic Ocean.

What Ocean east?
Pacific Ocean.

What Ocean south?
Indian Ocean.

What Grand Division west?
Europe.

What Grand Division southwest?
Africa.

What Sea and Bay south?
Arabian Sea and Bay of Bengal.

What four Seas east?
Bering, Okhotsk, Japan, and
Yellow.

What Sea southeast?
China Sea.

LESSON LVI.

What Sea between Arabia and Africa?

Which is the largest Sea west of Asia?

What two Seas between Asia and Europe?

What Sea east of the Caspian Sea?

What large Country in the northern part of
Asia?

To what Empire does Siberia belong?

What Empire south of Siberia?

What Country in the southeastern part of the
Chinese Empire?

What Country west of Chinese Empire?

What two Countries of Asia furthest west?

What Country between the Arabian Sea and
Bay of Bengal?

What two Countries northwest of Hindostan?

Between what two Countries is the Persian
Gulf?

What Mountains between Chinese Empire and
Siberia? Between Hindostan and Thibet?
Between Asia and Europe?

THE GREAT WALL OF CHINA was built more than two thousand years ago, to protect the Chinese from their enemies on the north. It extends over hills and plains, is about thirty feet high, and so broad that six horses can walk abreast on the top of it. Its length is 1,500 miles, or about the distance between Maine and Texas. It is strengthened by large square towers.

What can you say of Asia?

Asia is the largest and first inhabited Grand Division of the Earth.

Who were our first parents?

Adam and Eve, who lived in Asia.

Where was our Saviour born?

In the western part of Asia.

Of what does the Empire of Japan consist?

Of Islands, the largest of which is Hondo.

What articles come from Asia?

Furs from Siberia, Tea from China, and Coffee from Arabia and Java.

This is a picture of a HEATHEN TEMPLE or place of worship. It contains frightful looking objects, before which people fall on their knees and faces and pray. They are IDOLS, or false gods, which these people worship. They are made chiefly of stone or wood. Such people are called IDOLATERS, PAGANS, or HEATHENS. They believe these idols can hear their prayers, and grant what they ask.

You will be surprised to learn that there are millions of idolaters. They live in Asia, Africa, and the islands of the Pacific Ocean.

Missionaries have been sent from the United States and Europe to teach these ignorant people about the TRUE GOD who says, in his commandments, "THOU SHALT HAVE NO OTHER GODS BUT ME."

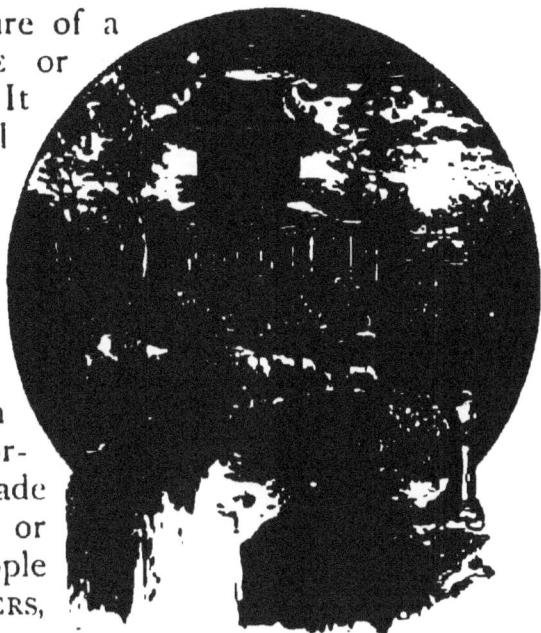

CHINESE GATHERING TEA.

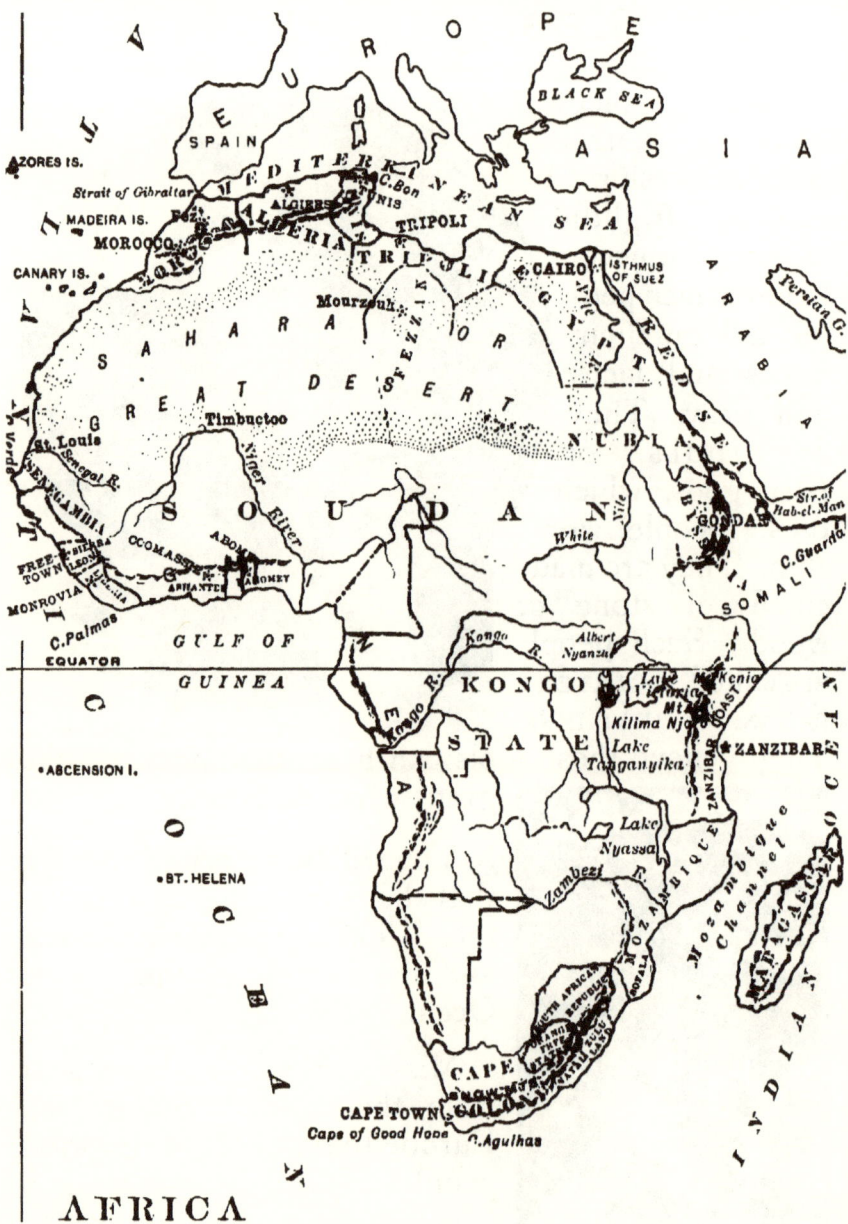

AFRICA

What Division of land is Africa?

A Peninsula.

Between what two Oceans is Africa?

Atlantic and Indian.

What Division of the Earth north of Africa?

Europe.

What Sea north of Africa?

Mediterranean Sea.

What Sea northeast of Africa?

Red Sea.

What Division of the Earth?

Asia.

What Isthmus between Africa and Asia?

The Isthmus of Suez.

What Gulf west of Africa?

Gulf of Guinea.

What large Island southeast of Africa?

Madagascar.

What Cape in the south?

Cape of Good Hope.

LESSON LX.

Which are the largest Rivers in Africa?

What Plain in the north?

What Region south of the Great Desert?

What Country west of Soudan?

What Region east of Guinea?

What Mountains in the eastern part of Africa?

What Lake in Soudan?

What Cities in Northern Africa?

With what Ocean is the Mediterranean Sea connected?

With what Ocean is the Red Sea connected?

What Countries of Africa touch the Red Sea?

What large River flows through them into the Mediterranean Sea?

What large River flows into the Gulf of Guinea?

On which side of Africa is Guinea?

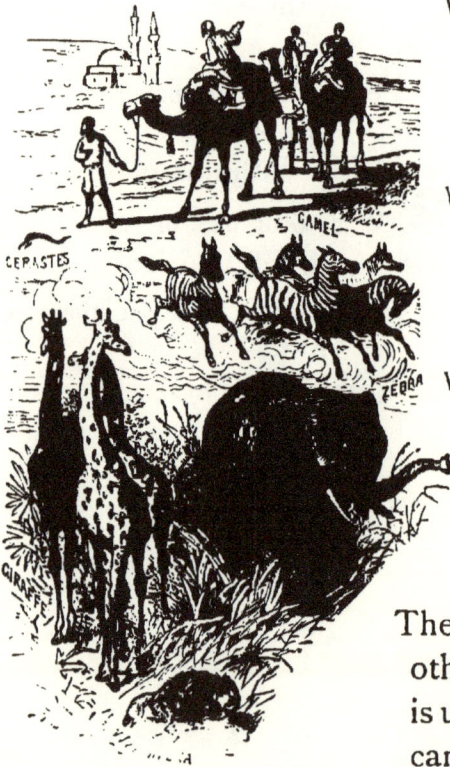

What can you say of Africa?

It is the hottest Grand Division of the Earth.

What animals in Africa?

The Elephant, Lion, Leopard, Hyena, Zebra, and others.

What dangerous Reptiles in Africa?

Crocodiles and Serpents.

What Trees abound in the forests?

The Cocoa-nut, Date, and other Palm-trees. Palm-oil is used in making soap and candles.

What storms sometimes overtake Caravans in the Great Desert?

Storms of scorching sand, raised by the wind.

A SAND-STORM. Men and camels must lie on the ground till it is over.

Africa is noted for its extreme heat, ferocious animals, and Great Desert.

The Great Desert is about 3,000 miles long, and 1,000 miles wide. It contains vast, sandy plains which are dry, hot, and barren, except in some green places, called Oases. People cross the desert in large companies, called Cara-vans, in order to defend themselves from robbers. Camels are used in crossing the desert.

Lions, elephants, and other wild animals are found in Africa as well as in Asia.

RUINS AND PALM-TREES IN EGYPT.

Elephants are hunted for the ivory, of which their tusks are composed. Many have been tamed, and are very gentle.

COFFEE.

TEA.

The **Tea Plant** grows to the height of five or six feet, and is cultivated in China and Japan. The leaves are gathered when green, and dried on heated pans. The color of green tea is due to a coloring matter that is dusted over it in the pans.

The branches of the **Coffee Tree** are loaded with berries, which look like red cherries. Each berry contains two grains or seeds of a light green color, which resemble beans cut into halves. These are roasted and ground before the coffee is ready for use.

Coffee comes from Arabia, Java, South America, and the West Indies.

WHEAT.

OATS.

Wheat, Rye and **Oats** grow on the top of the plants or stalks. When ripe they are cut something like grass; then the grain is removed from the husk, by being thrashed or beaten. The stems or stalks we call *straw*. Wheat is ground into flour, oats into oatmeal.

INDIAN CORN.

Indian Corn, or **Maize,** grows upon a stalk higher and thicker than that of the other grains. It is found in ears on the plant. When ripe and dry, the grain is ground into *Indian,* or *corn meal.*

The Tobacco Plant, when fully grown, is cut, and hung up to dry. From the leaves are made smoking and chewing tobacco, cigars, and snuff. Tobacco was first used in America.

TOBACCO.

The **Cotton Plant** is extensively cultivated in our warm Southern States. It grows from seeds sown in the spring. In the autumn the soft, white substance called cotton is taken from the pod or shell and separated from the seeds inside. It is then ready to be spun into threads, and woven into muslin, etc.

Flax is a plant which has a slender stalk, and grows to the height of two or three feet. The skin or bark consists of fine fibers that may be separated and spun into thread, then woven or made into cloth, called *Linen, Cambric, Lawn Lace,* &c. The seeds yield an oil called *Linseed Oil.*

COTTON.

FLAX.

COUNTRIES.

Where situated?　Bound them.

United States?	Brazil?
Greenland?	Spain?
Russia?	Siberia?
China?	Mexico?
England?	Persia?
Venezuela?	Central America?
Hindostan?	German Empire?
Arabia?	Scotland?
Turkestan?	Ireland?
Austria?	Turkey?
France?	Argentine Republic?
British America?	Italy?
Morocco?	Egypt?

MOUNTAINS.

Where are they situated?

Rocky?	Blue Ridge?
Himalaya?	White?
Andes?	Altai?
Alleghany?	Pyrenees?
Sierra Nevada?	Ural?

CAPES.

Where are they?　Into what waters do they project?

Cod?	Farewell?
Good Hope?	Horn?
Hatteras?	St. Roque?

ISLANDS.

Where are they?　By what waters are they surrounded?

Greenland?	Iceland?
Australia?	Madagascar?
West Indies?	Japan Is.?
Cuba?	Sicily?
Tierra del Fuego?	Hawaiian?

SEAS, GULFS AND BAYS.

Where are they?　Into what waters do they open?

Mediterranean S.?	G. of Guinea?
G. of Mexico?	B. of Panama?
Arabian S.?	G. of California?
Hudson B.?	Yellow S.?
Black S.?	Bering S.?
China S.?	Baffin B.?
B. of Biscay?	Delaware B.?
S. of Japan?	G. of St. Lawrence?
White S.?	Baltic S.?

STRAITS.

Between what lands are they?　What waters do they connect?

Davis?	Hudson?
Magellan?	Bering?
Florida?	Gibraltar?

RIVERS.

Where do they rise?　What courses do they take?　Into what waters do they flow?

Amazon?	Mississippi?
Nile?	Delaware?
Missouri?	Rio Grande?
Susquehanna?	Ohio?
Genesee?	Hudson?
Mackenzie?	Savannah?
St. Lawrence?	La Plata?
Potomac?	Columbia?
Kennebec?	Tennessee?
Niger?	Cumberland?
Orinoco?	Danube?
Connecticut?	Arkansas?
Kongo?	Zambezi?

LAKES.

Where are they?　What are their outlets?

Superior?	Huron?
Great Bear?	Champlain?
Great Salt?	Maravi?
Ontario?	Michigan?
Great Slave?	Erie?

CITIES.

In what Countries or States are they?　On or near what waters?

London?	Rio Janeiro?
New York?	St. Louis?
Constantinople?	Galveston?
Mexico?	St. Paul?
Lima?	Atlanta?
Lisbon?	Montpelier?
Boston?	Harrisburg?
Washington?	Havana?
Paris?	Cincinnati?
Buffalo?	Charleston?
Montgomery?	Nashville?
Madrid?	San Francisco?
Dublin?	New Haven?
Albany?	Calcutta?
St. Petersburg?	Jefferson City?
Montreal?	Annapolis?
New Orleans?	Trenton?
Philadelphia?	Santa Fe?
Cairo?	Tallahassee?
Richmond?	Pittsburg?

Eclectic School Readings

A carefully graded collection of fresh, interesting, and instructive supplementary readings for young children. The books are well and copiously illustrated by the best artists, and handsomely bound in cloth.

Folk-Story Series

Lane's Stories for Children	$0 25
Baldwin's Fairy Stories and Fables	.35
Baldwin's Old Greek Stories	.45

Famous Story Series

Baldwin's Fifty Famous Stories Retold	.35
Baldwin's Old Stories of the East	.45
Defoe's Robinson Crusoe	.50
Clarke's Arabian Nights	.60

Historical Story Series

Eggleston's Stories of Great Americans	.40
Eggleston's Stories of American Life and Adventure	.50
Guerber's Story of the Chosen People	.60
Guerber's Story of the Greeks	.60
Guerber's Story of the Romans	.60
Guerber's Story of the English	.65
Clarke's Story of Troy	.60
Clarke's Story of Aeneas	.45
Clarke's Story of Caesar	.45

Natural History Series

Kelly's Short Stories of Our Shy Neighbors	.50
Dana's Plants and Their Children	.65

Copies of these books will be sent, prepaid, to any address on receipt of the price by the Publishers:

American Book Company

Language Lessons
AND
Elementary Composition

The following books are adapted for beginners in the study of Language and of Composition :

Long's New Language Exercises. Part I. . . . $0.20
New Language Exercises. Part II. . . .25
Lessons in English Grammar and Composition .35
Maxwell's First Book in English 40
Metcalf and Bright's Language Lessons. Part I. .35
Language Lessons. Part II. . . .
Metcalf's Elementary English 40
Swinton's Language Primer 28
Language Lessons 38
School Composition 32

LANGUAGE TABLETS AND BLANKS

National Language Tablets. 12 Nos. Per doz. .90
Stickney's Child's Book of Language. 4 Nos. Each .08
Letters and Lessons in Language. 4 Nos. Each .16
No. 5. Grammar 35
Ward's Grammar Blanks. 2 Nos. . Per doz. .90

These Tablets and Blanks supply a great variety of graded exercises in Language, Grammar and Composition for practice and review. Their use in classes will economize the time of both pupil and teacher.

Specimen copies of any of the above books will be sent, prepaid, to any address on receipt of the price by the Publishers:

American Book Company

New York   •   **Cincinnati**   •   **Chicago**

WEBSTER'S
SCHOOL DICTIONARIES

Revised Editions

Webster's Primary School Dictionary . .	$0.48
Webster's Common School Dictionary .	.72
Webster's High School Dictionary . .	.98
Webster's Academic Dictionary . .	1.50
Webster's Academic Dictionary, Indexed .	1.80

Webster's School Dictionaries in their revised form constitute a progressive series, carefully graded and specially adapted for Primary Schools, Common Schools, High Schools, Academies, etc. They have all been thoroughly revised, and made to conform in all essential points to the great standard authority—**Webster's International Dictionary.**

Webster's Dictionaries have been endorsed and recommended by the Executive Departments of the U. S. Government, the highest State Courts, the U. S. Government Printing Office, the U. S. Bureau of Education, the Educational Authorities of all the States, College Presidents and Faculties, Authors and Journalists, Teachers and Scholars.

The use of **Webster's School Dictionaries** should be made as general in schools as other text-books. No school is properly equipped until every desk is supplied with a School Dictionary suitable for the age of the pupil.

Copies of Webster's School Dictionaries will be sent, prepaid, to any address on receipt of the price by the Publishers :

American Book Company

New York • **Cincinnati** • **Chicago**

SUPPLEMENTARY READING

JOHONNOT'S HISTORICAL READERS

SIX BOOKS. 12MO. ILLUSTRATED

Grandfather's Stories. 140 pages	27 cents
Stories of Heroic Deeds. 150 pages	30 cents
Stories of Our Country. 207 pages	40 cents
Stories of Other Lands. 232 pages	40 cents
Stories of the Olden Time. 254 pages	54 cents
Ten Great Events in History. 264 pages	54 cent·

JOHONNOT'S NATURAL HISTORY READERS

SIX BOOKS. 12MO. ILLUSTRATED

Book of Cats and Dogs. 96 pages	17 cents
Friends in Feathers and Fur. 140 pages	30 cents
Neighbors with Wings and Fins. 229 pages . . .	40 cents
Some Curious Flyers, Creepers and Swimmers. 224 pages .	40 cents
Neighbors with Claws and Hoofs. 256 pages . .	54 cents
Glimpses of the Animate World. 414 pages . . . $1.00

These books are admirably adapted for use as supplementary readers. Each series contains a full course of graded lessons for reading upon instructive topics, written in a style that is of the most fascinating interest to children and young people, while training them to habits of observation and storing their minds with valuable information. Each book is fully illustrated in an artistic and attractive manner.

Copies of any of the above books will be sent prepaid to any address, on receipt of the price, by the Publishers:

American Book Company

www.ingramcontent.com/pod-product-compliance
Lightning Source LLC
Chambersburg PA
CBHW020239090426
42735CB00010B/1768